Let Your Ideas Speak Out

Guide to Preparing and Marketing Spoken-Word Audiotapes and CDs

By

Eugene D. Wheeler and Rennie Mau

Pathfinder Publishing
of California
Ventura, CA

Let Your Ideas Speak Out

Guide to Preparing and Marketing Spoken-Word Audiotapes and CDs

By
Eugene D. Wheeler & Rennie Mau

Published By:
Pathfinder Publishing of California
458 Dorothy Avenue
Ventura, CA 93003
(805) 642-9278

Library of Congress Cataloging-in-Publication Data

Wheeler, Eugene D.

Let your ideas speak out: guide to preparing and marketing spoken-word audio tapes and CDs / by Eugene D. Wheeler and Rennie Mau.

p. cm.

Includes index.

ISBN 0-934793-33-6 : $8.95

1. Sound recording industry--Management. 2. Sound recordings--Production and direction. 3. Audiotapes. 4. Compact discs.

I. Mau, Rennie, 1954- . II. Title.

HD9697.P562W44 1991

621.389'32--dc20 91-17889

CIP

ISBN 0-934793-33-6

CONTENTS

Acknowledgements

The authors would like to thank the following people:

Rennie Mau: Linda, Danielle, and Jennifer for their patience, love and support. And to my good friends Richard, Denis, Brad, and Joy.

Eugene D. Wheeler: Many thanks to Dorothy Towvim for reviewing the manuscript; Lou Hartney for her proofreading and review; Jan Nathan for listening to the idea and suggesting Rennie; Kathleen Sublette for her comments and suggestions; Dave Fischer for his cover desgn; Dan Poynter for his review and helpful suggestions; and to Eugenie Wheeler for her editing, proofreading, and continuing support.

INTRODUCTION

This book has been written to give writers, educators, trainers, print publishers, and multi-media publishers a guide to preparing for a growing need and market: translating ideas normally found in printed form into the spoken-word format and onto audio tapes, CD discs, and the new multimedia technologies. Audio tapes have been around a long time, but have been overshadowed by videotapes since the early 1970s. The first books-on-tape were released by major publishing houses only around 1985, but smaller independent publishers produced them much earlier.

While there are high quality audio tapes on the market, many are of mediocre quality. Most don't survive after the first year of competition. The audio tape and CD market is growing, but so is the competition. This book will provide you with information on the audio recording process that a potential audio tape or CD author or publisher should have. It offers suggestions about how to make your product better and how to find a market for it.

1

THE EXPANDING WORLD OF AUDIOTAPES AND CDs

Most writers and producers/publishers are interested in making an audiotape of their written words for three major reasons. First, they want to reach their target population with their particular message. They want to inform, educate, inspire, entertain, or motivate their target audience. A second reason is to influence a wider range of people with their ideas. A third is to expand the economic well-being of their company or in other words, make more money. The spoken-word audio cassette industry has only been around for thirty years. Prior to this the spoken word was captured on phonograph records. The industry is presently highly competitive and secretive.

Is there room "out there" in the marketplace for more audiotapes, and for yours in particular? The answer is YES. There is room "out there" for more audiotapes. But the market for your particular tape will depend on how good your ideas are, the quality of the tape production, and how aggressive and thorough you are in marketing your product.

A Little History of the Audio World

To better understand the audio potential let's take a short timeout and look at what has been happening in the audio field. Understanding trends can be useful in marketing your products. Besides, a little perspective never hurt anybody.

Media Literacy in the 1990s

Studies show that a person in a conversation uses between 4,000 to 5,000 words an hour. When a conversation has fewer pauses, such as in a radio talk show, as many as 8,000 to 9,000 words an hour are used. In reading, a person at a normal pace can read about 14,000 to 15,000 words per hour. In an average day a person can use as many as 100,000 words in the four basic processes of communication: reading, listening, talking, and writing.

Importance of Sound in Our Culture

We live at a time where the visual image (film, video) has become an important influence on our lives. Today, the average person spends 4-6 hours a day in front of a video screen. And this media is less than a hundred years old. Sound plays an important role in viewing movies and visuals. Most people don't realize that without sound many of these images would not make sense. A person listening to a TV soundtrack without observing the visual picture can understand most of what is happening on the tube, but a person observing the picture without sound will probably not understand what is occurring.

In this decade, as companies and investors pour millions of dollars into research and development to produce computers and hardware for the promises of "multimedia," it is amazing how many of these same developers do not understand that 50% of multimedia is sound. Many of these people have no idea as to the importance of sound and human perceptions of listening. The spoken word has been around since humans first learned to communicate with mouth sounds. The individual human mind acquires, comprehends, produces, and stores language with the internal spoken word.

Printing has been around less than a thousand years. Before Gutenberg, humans lived in a primarily speaking and listening culture. With Gutenberg, a romanticized culture developed as well as typography. Prior to print, stories were told through oral symbols. In recent decades a bias has evolved in the Western World against oral traditions and cultures. But how do you represent a flow of speech on paper? The printed word cannot do this. For centuries scholars read their work out loud. Until the 19th Century, the Christian bible was read out loud, too. Without getting too technical or into linguistics, the human ear when, "listening," perceives the uniqueness of language: word rhythms, sound patterns, metrical phonology, sound combinations, and stressed syllables.

While educators complain about low reading skills in schools, they fail to realize they have not emphasized good listening skills. Hearing is not the same as listening. Students who read, read externally. Hence we have a nation with a sixth grade reading proficiency. Studies show that children today listen to only 25% of what they hear. In the average school today students are taught to read to themselves. Reading periods are quiet times. Few read out loud in class and when they do, their vocal qualities and gestures become constricted (as opposed to when they are talking in normal conversation with another person). Unfortunately, due to lack of practice and training, when many people today read out loud, they produce a voice sound that is a monotonous drone.

Re-Emergence of the Audio Cassette

In the last few decades the video media have attracted the public's attention. Educators put audio cassettes on hold. People wanted to see, not hear, what was of interest to them. Marketers de-emphasized the audio cassette. It was thought to be only for blind people or children "read-alongs." Meanwhile, audiotape producing companies such as Books-On-Tape, Listening Library, HighBridge, Dove, and others, did a booming business in producing and selling tapes.

In the thirty years that the compact audiotape has been around, it was utilized minimally. However, in the 1980s the personal portable cassette (e.g., SONY WALKMAN ®) became a necessity and a way of life. Walkers, hikers, commuters, workers, joggers and many others used them daily. Today, the average American home has 3.5 audio cassette players.

Political Use of Recorded Sound

The use of recorded sound has helped to shape the geopolitical arena by friends and foes alike. Radio Free Europe, the Voice of America, and the BBC have spread the message of hope and encouragement to people whose access to information was limited by the controlled media in their countries. Missionaries have used recorded material to communicate with cultures unable to adapt to a new written language. Even today, village people gather around hand-cranked audio players. Dictators such as Hitler and the Ayotollah Khomeini understood the power of the spoken-word and used it to communicate their thoughts to their followers.

When Desert Storm began in the Gulf it was amazing to see how many of the video networks were forced to rely on the spoken word. Also many of the radio stations were playing live feeds from the television stations. One could listen and know what was going on. Millions listened to Saddam Hussein's speeches during the Gulf War while the television camera concentrated on a stationary radio.

Existing and Potential Market for Audiotapes

National Growth and Use of Audio Cassette Players.

According to Bowker's *On Cassette: A Comprehensive Bibliography of Spoken-Word Audio Cassettes in 1990,* approximately 4,500 audiobook tapes were released in the U. S., raising the total number of audiobooks to 38,000. This is three times the number of audio cassettes in circulation five years before.

New uses for audiotapes are being introduced. Some producers see a trend toward magazine/newsletter tapes, which run about 60 minutes each. Sales and management tapes, religious tapes, seminar tapes, and how-to tapes, such as "how to buy a house," are all doing well. In general, the major audio market appears to be open to print publishers in the fields of:

- Self-Improvement—How to lose weight, change your personality, control drinking, stop smoking, and enhance your life.
- How-To Books—How to play better golf, tennis, make a sculpture, handle animals, etc.
- Educational Books—Includes elementary/high school and the college market. This includes institutions and can be a big market. Tapes in this genre include remedial and developmental learning.
- Government and Industry Training Programs—This is a big area covering human relations, data processing, computer use, management, and various technical subjects.
- Seminars and Workshops—Tapes of seminars and workshops held at conferences of professional, educational, and self-help conferences. It can be cheaper and more effective to buy a tape of a workshop rather than attend it in some city miles away.

Various distributors and producers caution that the quality of tapes is important to their survival.

Growing Audiotape Market

The existing market for audiotapes continues to grow. In 1990, the American Booksellers Association had 150 exhibitors for audio. Distributors are now more open to audiotape sales. Some distributors specialize in them. Grammy Awards now have an audiobook category. Many audio publishers are beginning to face stiff competition from large trade houses that have jumped on the audio bandwagon. There has been a threefold increase since 1985 of audio publishers, from 300 to nearly 1,000 in 1991.

Print publishers are beginning to understand how the spoken-word audiobook, rather than competing with their products, enhances and complements them. Simultaneous releasing of audiobooks in the last five years has increased dramatically.

From 1985 to 1990 there was a 300% increase in audio cassette duplication. Some of the many uses now being made of the simple audiotape are:

- Audio calling cards/promotion
- Audio tutorials
- Audio guides
- Self-help
- Read-alongs
- Education
- Audio books —fiction and non-fiction
- Direct mail promos
- Audio newsletters
- Audio journals/magazines
- Demonstrations (products/services)
- Exhibit presentations
- Marketing
- Radio adaSpots
- Radio dramas
- Training
- Audio for video
- Voice-overs for animation
- Voice-overs for documentaries

Ingram Audio Company, a division of the largest book wholesaler in the world, and an audio distributor to bookstores, lists 20 different audio categories which are:

- Business-Professional
- Children's Activity
- Children's Classics
- Children's Music
- Children's Educational
- Drama
- Exercise
- Fiction-Abridgements
- Fiction-Complete
- Humor
- Inspirational/Bibles
- Language
- Nonfiction
- Poetry
- Radio Shows/Sound Tracks
- Relaxation/Mediation
- Self-Help
- Self-Hypnosis/subliminal
- Travel

These categories will give you an idea of how Ingram sees the audio potential. Ingram also publishes, *Audio Advance,* a preview newsletter of new audio products.

Trends and the Future

Many computer companies have bundled audio cassettes with their computer software. Companies such as Personal Training have done a booming business in audio cassette tutorials. Separate technologies are now combining medias into one. Companies using terms or names like "multi-media" have evolved in the business. An understanding of how to use the newly created converging medias enables humans to cross media bridges.

In the 1990s, organizations began forming joint ventures of the two major disciplines: video and audio publishers. For instance, Lucas Films, known for its great movies, is also known for its outstanding sound studios and sound libraries. It is anticipated that computer software publishers in the 1990s will begin to have joint ventures with print publishers, video publishers, and audio publishers. Many will become multimedia publishers. Educators and businesses are also beginning to harness the new media to communicate their information. Museums are making greater use of audio by creating interactive "soundscape" environments in halls and exhibits. Airlines have specialized programs in spoken-word magazine formats. Exhibitors are beginning to realize the importance of sound along with the visuals. Some exhibitor vendors actually include speakers in walls of their booths.

The Audio Process

The audio process of creating and recording sounds, whether the sound is put on tape or CDs, has not changed in decades. While recording engineers debate whether to utilize analog or digital technologies, the "audio script" remains the same. It is formed by voice talent and sound creation.

Sales from spoken-word audio books are on an upswing. While audio books may never rival a printed book, they can add substantial income to a publisher. The first audiotape books were produced by entrepreneurs—not by the big publishing houses. That has changed. Big publishers are now in the spoken-word tape business. An early pioneer was Random

House who formed the Random House Audio publishing division. They have been at it for six years and produce approximately 40 to 50 titles a year. Others in the field are Bantam Books, Harper & Row and Simon & Schuster. Random House Audio was one of the first to release a tape and a book simultaneously.

Many of the novels produced on tape are abridged. For instance, Gore Vidal's 496-page book, *Hollywood,* sold for $19.95, and the abridged three-hour tape sold for $14.95.

Lynda Sheldon of Random House Audio indicated in an article by Tom Blackburn of the Cox News Service, that tape sales are a long way from rivaling books. If a book sells 300,000 to 500,000 copies, best-selling numbers, the best-selling tape will sell 35,000 to 50,000 copies.

It wasn't until September of 1988 that Publishers Weekly started regular reviews of tapes. Now a tape review section is a regular feature. As Publishers Weekly stated: "Since the section was launched, we've been increasingly deluged with titles, testimonials to the aggressive growth and diversity of the industry."

Tapes are now available for all age groups; preschool to about age 12, teen-agers, adults and special interests. Just as women tend to buy more books than men, in the audiotape industry women are discovering tapes, and sales are increasing to that market.

Since the introduction of CDs a few years ago, people expected the demise of the audiotape. But it didn't happen. In fact, from 1986 to 1990 there has been an increase worldwide of audiotape duplication. In 1991, a new digital technology was introduced into duplication houses that makes the average quality of an audiotape equal to that of a CD.

To provide a sales outlet for spoken-word publishers, more audiotape and CD catalog services are now available. A few notables are Chinaberry Book Service, Spring Valley, CA; Music for Little People, Redway, CA; High Windy Audio, Fairview, N.C.; The National Catalog of Storytelling, National Storytelling Resource Center, Jonesborough, TN; B. Daltons, and Waldenbooks.

Rental Outlets

Video rental outlets are now marketing audiotapes for rent. Specialized mail order rental catalogs are being developed to sell tapes. Village Story Tapes of Sebastopol, CA, had a sales growth rate of 500% in one year.

Advantages and Disadvantages of Audiotapes and CDs

Audiotapes have many advantages. A major reason is that they are an inexpensive form of entertainment and learning. Audio presents stimulating verbal messages that are much more dramatic than similar messages in print. This medium is very portable as the tape players are small and light. The tapes are durable and easy to use. A tape recorder/player is easier to use than programming a VCR.

Little storage space is needed to store tapes. The tapes can be listened to almost anywhere or anytime; driving, walking, jogging, exercising, doing office or housework, waiting, or in classes. The auto commuter can use transit time more beneficially by listening to educational tapes; self-help, how-to, novels, or foreign language tapes. Tape players are compact and can fit into a pocket or purse easily. With tape you can pace your own learning of a subject of interest. The tapes can be reproduced quickly and inexpensively.

The limitations of using audiotapes are few. As the audio media fixes the sequence of a presentation, you must rewind or fast forward the tape to hear selected material over again. With CD disks, based on a digital system, the listener can go directly to the desired section.

Another limitation is that a tape player needs to be cleaned and maintained periodically to ensure good fidelity of the sound. If cheap tape is used, occasionally the tapes will stick or become tangled. If a tape is broken in a cassette, it is not easy to repair unless you have a repair kit.

CDs have better sound qualities, can be programmed quickly, and desired sections can be selected and found easily. There is also little wear and tear on a CD because of the laser technology. No needle or playback head touches the disk. However, at the present time, the costs of CDs are considerably higher, but are dropping. CD players are also expensive, but are decreasing in price.

Multimedia

At present, CDs have been mainly CD ROMs, playback only. But now CD-I (CD Interactive) and new recordable media CDs will become standard. In the 1990s, publishing will converge technologies. Perhaps the most revolutionary development to transform the concept of a book in print to a new medium will be the Sony DiskMan, a portable CD player that will display text as well as play back spoken-word recordings.

The process by which an audio script is produced is basically the same as for a multimedia product. Multimedia producers can really bring their production costs down by sequencing text, graphics, and moving images to an audio script/track. In early animation, the soundtrack was produced first and then the image was sequenced to it. Audio producers have found that many multimedia hardware and software companies touting the new technologies have placed the cart before the horse by creating text, graphics, and moving images first and then trying to add the audio tracks. It may be easy sequencing music, but sequencing a narration to material after the fact is cost intensive.

One company that spent several hundred thousand dollars producing interactive training CDs discovered to their horror that they had considered the audio track last. It cost them a large sum of money to hurriedly hire narrators and then sequence music and sound effects to the already created moving images, text, and graphics.

Because a person knows how to use a word processor does not make him a writer. It is the same for multimedia hardware and audio software. Just because one has the capability of recording and editing sound on a computer does not make that person an experienced and trained sound engineer.

NOTES

2

QUESTIONS TO ASK IN PREPARATION

What Material Is Good for Audiotapes and CDs?

Most types of fiction and non-fiction are good for audio with the exceptions of highly technical books and art books that have many illustrations. Self-help books have been sold successfully as audiotapes, but usually have a life of about two years in bookstores. However, an audio book can have a longer life if it provides needed information, entertainment and is marketed well. According to Jim Brannigan of HighBridge Productions, their tape of Garrison Keillor's, *Lake Wobegon Days,* has been a best seller since 1982. To date, novel-audio books tend to be more popular and sell better than non-fiction spoken-word books. But this pattern could change as publishers and book stores change policies.

Some audio cassettes made from movie sound tracks are having great success. When the movie *The Little Mermaid,* first came out critics made light of it. But the cassette that was developed from the sound track sold over a million copies in a short period of time after its release. Guess who laughed all the way to the bank?

There is some debate in the industry as to whether a child's tape should use adult voices or children's voices. Children and youth today are very sophisticated and audio literate. Many children don't like audio books when the voices of the children on the tape have been made by adults. Apparently this bores them. Some new children's books are in 3D sound, which intrigues them, and are successful.

First-time producers sometimes make the mistake of thinking that children's audiotapes are simple to produce. In reality they are harder to produce than adult material.

How Should You Format Your Material On Tape?

The format is important. Should it be in first person, second person, or third person? Do you want the reader to have a formal voice or a warm friendly voice to relate intimately to the listener? It is important to think of the final results you want on your audiotape or CD before contacting an actor or friend to read any voice parts.

Where Do You Go For Audio Production Help?

Probably the first place to go is to a friend who has had experience making an audiotape. If you don't know anyone with experience, the next best step is to contact a reliable audio production house or audio studio.

Decisions, Decisions

Before you go to a studio or to a media production house to record your material for audio, be prepared to answer many questions concerning the recording, production, duplicating and packaging of your audio book or message. It is a good idea to know the vocabulary of the business so that you sound knowledgeable and don't come across as a "tenderfoot."

You will be asked: How many pages are in your script? Desired time of tape? Any sound effects? How many needle drops? Is the script formatted and marked for reading?

You will also be asked questions like: What quality tape? How many Working Masters? How many minutes will the tape run? What kind of tape cassette, five screw or welded? How many Norelco boxes and what color? The more you learn about the process and are aware of the decisions that need to be made, the more smoothly the project will flow and the fewer mistakes will be made. Once you get the hang of the vocabulary and procedures, the process won't seem too complicated. A glossary of terms is included in the Appendix to help you understand the terms. **So let's get started!**

Major Decisions to be Made by You as the Producer/Publisher of an Audiotape.

- What are your goals for your tape?
- How many tapes do you hope to market?
- Who will prepare the audio script of your material or book? The author or an experienced scriptwriter?
- How long in minutes will your tape run? Will you need one, two, three or more tape cassettes?
- Who should be the reading voice? The author, or a professional actor? Many audio book production companies stress having actors or a reader with acting ability rather than just voice-over specialists.
- Are you going to use a voice coach? If so, who?
- Should you use music on the tape? If so, will it be original music or needle drop music?
- Should you use sound effects?
- Who will handle the recording and production of your tape? Will it be in-house or under contract with a media house or other professional?
- How much can you afford to budget for the project? Can you afford sound effects and music?
- How will you promote and market your project?

The above questions and many more must be addressed before you get into actual production. The remaining part of this book will address the above questions and help you learn how to get the answers.

3

WHEN YOU'RE READY TO PROCEED

PRE-PRODUCTION

Determine Goals, Scope, and Budget of Project

It is not only important that you determine the goal, purpose and budget for your audio project, but also that you write them down in a form that you and your staff can refer to in order to ensure that you are on track.

Forms are included in the Appendix as examples for you to use. To get started, try answering the following list of questions. First let's look at what the end product is for a publisher, for example an audiotape book.

What Is an Audiotape Package?

An audiotape album or package contains one or two tapes that are enclosed in a cardboard, plastic, or vinyl molded enclosure. Sometimes the cassette is enclosed in a plastic box, called a Norelco box. This box is sealed into a book-like container or case made of plastic or cardboard that has written material on both sides. The inside cover is much like a book. This package is shrink wrapped to protect the cassette from prying hands.

The average audio book consists of one tape, approximately 45 minutes on each side for a total of 90 minutes. The number of minutes or number of tapes depends on the material and what the author and producer/publisher believe should be included to cover the subject for the targeted audience.

Below, four examples are described to give you an idea of how to start setting goals for your project.

What Are The Project Goals?

Example: In the case of a publisher using a non-fiction book named *Quest For Respect*, the goal of the author and publisher is to reach and help victims of rape to overcome their traumatic experience. It is also to help rape counselors be more sensitive to the needs of rape survivors.

What Is the Financial Goal?

To sell 4,000 two-tape cassette albums at $15.95 (average net sale $12.00 or $48,000) in a two-year period, to make a profit (break even point is $11,000), and to provide royalties to the author.

Where Will the Audio Production Be Used?

To be used by rape survivors and counselors on personal tape players, in classroom settings, and in counselling sessions. It is anticipated that it will be played in rape crises centers, cars, Walkman ® sets, and in small conference rooms.

Is the Proposed Tape Project New or a Revision of an Existing Work?

It is a new tape project of an existing book.

Analyze Your Audio Script and Recording Needs

At an early stage you should determine who is to create and format the script for voice, and whether to use music and/or sound effects. If the material is to be abridged, shortened or changed, a script will be necessary. This calls for a scriptwriter if the author can't do it. You also must determine your cast needs. If you are to have several voices, it will be necessary to audition the cast for the various roles, unless you know who you want. If a voice coach is necessary, you must hire one. Rehearsals must be arranged for the cast of readers. If songwriters, composers, and arrangers are necessary, they must be recruited and hired.

Print or Script Editor

The script editor is important in preparing your material for reading for a recording, making it tight, easy to read, and editing it down to the desirable time span. A script editor first goes through the printed material to edit it down in time. On the second reading of the script, the script editor edits for language quality and flow.

In preparing a script for reading, keep in mind that the text on an eight and a half by eleven inch page, double spaced, will take an average of two minutes to read. To read an unabridged page from printed material, it is advisable to photocopy and enlarge the pages for easier reading. You can use that figure as a guide in determining how many minutes your material will take on an audiotape.

Writing and Editing for the Ear

To prepare a quality script for an audio reading requires special skill. An experienced editor or scriptwriter will know which words portray vivid images in the mind of a listener. Words generate visions and have unique sounds. Readers should always read their material aloud to spot any awkward sentences or words that don't fit well and to check for rhythm, pitch, and pace. It's important to ensure that the words and sentences flow smoothly when read out loud. A good sound editor can help make an ordinary tape recording much better.

Script Editor Costs

These editors can range from $15 to $50 per hour depending on the geographic area and the person's experience. Simple editing, without many changes, can cost as low as $15 per hour. If the project is more complicated, it can run about $50 per hour.

Reader's Role

The voice readers should read through the prepared script in advance and time themselves before the actual recording takes place. It is a good idea for readers to actually record themselves on a small tape recorder to check their pacing, their voice inflections, etc. This is especially important for the author-reader to do when inexperienced at recording in a studio environment.

Budget For Readers

If more than one or two readers are involved, remember to put funds in the budget for them if they are not volunteers. If your budget is tight and you can't afford a professional reader, do the best that you can without one. But get your most talented "friend."

Using Friends or Volunteers

Beware. Publishers often make the mistake of using friends to do the reading. While friends' voices may sound well in person, after an hour or so of recording, they may tire and their voices could become monotonous and boring to the listener.

Readers and Voice Coaches

Some authors are poor readers, while others can be very skilled. If an author is used, consider having a professional voice coach listen and guide him through the recording to ensure that the quality is good. A professional voice coach can gently advise a reader in such a way that it keeps the reader calm. Sometimes it is advisable to have a voice coach assist the reader prior to the recording session. Producers often serve as voice coaches. If a reader should become angry or tired, his voice will show it. A clear, calm voice is important to obtain a quality recording.

It may bother some readers to be put in a room by themselves with people watching from behind a glass-enclosed sound booth. Sometimes the cold and formal studio environment will make them nervous. A skilled producer or voice coach can put the reader at ease which will be reflected in a relaxed voice recording. It is better if the coach sits with the reader rather than in the sound booth. A good voice coach can make an inexperienced reader sound like an expert. Children especially, are intimidated by studios and reflect it in nervous or high pitched voices. A voice coach can also make certain that questions or emotive phrases are emphasized so that the listener hears them. The coach can observe and prevent the reader from moving too close or too far from the microphone.

Cost of a Voice Coach

While costs will vary in different parts of the country, the range in California is from $15 to $50 per hour. Hiring a high school or college voice teacher can sometimes help the producer. Allow an hour or so for the coach to become familiar with the script before the recording actually starts. Budgeting a few hours of time for a voice coach can make the difference between a mediocre recording and a great one. Even some professional readers use a voice coach during recordings to monitor the quality of the voice.

Use of Music and Sound Effects

If you can afford it, add some music and/or sound effects (SFX) to your recording to enhance effectiveness and salability. Many publishers doing audio books like to include music in the introduction as a bridge between sections, at the end of each side of the cassette, and at the introduction of each side. You can obtain music in two ways: by way of a "needle drop" or by using original music. Needle drops are pre-recorded music that you insert in the location desired. The publisher pays a fee for the use of the music. To obtain original music you must hire a composer and pay musicians to record it and put it in the desired spot on the tape.

Some audio companies add an "audio logo" to the beginning of the tape to identify the publisher. Some companies state the copyright year and that all rights are reserved.

Songwriter's Role

A songwriter can be paid by fee or on a royalty basis. When you're having original music composed, a contract and release should be prepared and signed before proceeding with your project. See sample Work For Hire contract in Appendix, page 71.

Cost of Songwriters

A songwriter can cost as low $200 to $500 per original song, or can go much higher for well-known writers. Most producers have a "buy out" agreement that entitles them to own all rights to the song. Studios usually know the "thumb print" or the type of music written by local musicians.

The Arranger's Role

An arranger takes the music already composed and arranges it for the tape. The music can be performed by a synthesizer or a band. An arranger can receive from $400 to $500 or more for getting the song and band together.

Use of Sound Libraries

Sound libraries are also a source of music for you. A variety of sound libraries are available on cassettes. Libraries include sound effects and music. A good recording studio has its own sound library and is sometimes licensed to use material or compositions. The staff will know where music can be obtained or who is available to write original music.

Permissions and Releases

You must obtain written permission from each reader, singer, actor, or musician used in the production to use his or her services on your audio cassette tape. This also applies to certain authors of quotes recorded on the tape and to anyone used, whether a volunteer or a paid person. The release form should include using the voice for radio, cable TV, satellite, or any medium that the project might be converted to at a later time.

There have been cases when an artist has sued for royalties when the services were used on a different medium than originally was produced. A sample release form is included in the Appendix, page 72, that gives you an example of what to include in your release form.

The Peggy Lee Case

In 1991, singing star Peggy Lee, at 70 years of age, sued the Walt Disney Company for $50 million for a breach of contract and unlawful enrichment over a video cassette release of *Lady and the Tramp*. Lee had co-written six songs and performed the voice of four characters, a dog, two cats and Lady's owner, Darling, in the 1952 film. The movie earned more than $140 million, including $90 million from the video tape sale alone. At the time Lee participated in the film she was paid $3,500 for the use of her voice. The contract she originally signed did not anticipate video and didn't mention it.

In March of 1991, the jury in the case awarded Peggy Lee $3.8 million. The award included $2.3 million for breach of contract, plus $500,000 for unjust enrichment, $600,000 for illegal use of Lee's voice and $400,000 for the use of her name. **So be careful.**

If you don't obtain a release and your audio should become successful, your friendly artists might get a lawyer and sue you. This is also important in case you sell your product later to another company. If there is no signed release available, they might say that they won't take a chance on buying your product.

In another example, an audio publisher did a project with a number of children's voices twelve years ago. The person sold the audiotape to another company and the tape was marketed successfully. When the tape became popular, the lead singer, now twelve years older, sued for a considerable amount of money for her singing. To settle the suit, the company paid her off because she was on most of the songs on the tape.

Visual Release

Get a visual release at the same time you get a reader's release for audio. You might use the audio voices or music, etc., at a later time for a video production. The permission process is for your protection as well as for the protection of the actors, artists, and the producers.

License to Use Music

Keep in mind that you must have "permission" or a license to use someone's recorded music unless you create your own. You must obtain a Mechanical Royalty Release from the agent of the composer or the producer holding the copyright of the music or song you intend to use.

Work for Hire

The producer or publisher will handle a Work For Hire agreement with song writers and/or arranger. This agreement gives the publisher the ownership of their creative product.

Synchronization Release

This is obtained when you use someone's music with visuals or background to voices—you synchronize a picture to a sound or sound to the picture. This type of permission can be obtained by paying a fee up front or a royalty for each cassette or CD sold.

Determine if the production will be linked with any other published works, print, audio, computer, or video. Also, determine if the production will be used for radio spots, or as the audio track for future computer/video production.

Anticipate How the Audio Production Will Be Played

Where will your tape or CD be heard? Will it be listened to on home stereos, in automobiles, on portable Walkman-like players, in conference rooms, auditoriums, or classrooms? While this may be difficult to determine initially, it is important for selecting the grade or type of audiotape and recording process to use.

Determine Whether the Production Will Be or Include:

- Voice only?
- Voice and limited music background?
- Voice, limited music, and limited sound effects?
- Voice and fully produced music background?
- Voice, fully produced music, and full sound effects?

A voice-only tape will be a lot cheaper to produce, whereas one with voice, music, and sound effects will cost considerably more. More on this later.

Selecting a Place to Record Your Material

Should you record in a professional voice or sound studio, in-house, or in a media-House? Checking out a suitable place to do your recording is important. You are wise to go to a professional voice studio, not a music studio. But which one?

Selecting and Planning Time for a Studio

When arranging for time in an audio or recording studio, estimate one hour per 15 pages of double-spaced script for recording. Depending on the reader, add an extra hour or two in case it takes longer. So, reserve an extra hour or more at the recording studio. If you don't finish in the time allocated, you may have to come back at a later time because another party may have already booked the facilities at the end of your time. The studio has no alternative but to stop your production and go on to their next client. This could mean that your voice actors, musicians, voice coaches, etc., have to come back at a later time when the studio and your group are available. This can easily double your costs.

You should visit the potential studio ahead of time to check on the equipment available, listen to the sound of the room,and meet the production crew. Does it have "dead" sound which is good, or echoes, or a sound not desirable for your type of production? An echo sound can always be added if required in post-production.

Budgeting Time and Funds for Recording and Editing

Time from Concept to Completion of Audio Production

When you prepare an audio production plan and budget, put the various components of it into a Time Plan. For instance, to prepare 60 pages of printed material into a two-cassette audiotape or a CD, either the author or a scriptwriter should go through the book, mark the passages, and indicate what is to be dropped in order to come up with a two-hour tape. Approximately one to two weeks should be allocated for preparing and reviewing a script, depending on the length and complexity. Two weeks is needed for production (recording and editing) and two to three weeks for duplicating and packaging. Altogether, six to eight weeks should be allocated for preparing a print book in audio form. This includes organizing the actors, hiring the producer, voice coach, and other people involved, booking a studio, recording the material, editing and doing the duplicating.

For instance, it might take you only four hours to record your material in a studio, but it may take four to ten hours per recorded hour to edit and complete the master tape. In addition, add time for recording music and sound effects. While this is under way, the publisher should be working with the album designer and printer to ensure that the package will be ready when the tape is completed and duplicated.

Duplication Time

When your material has been recorded on an original master, a working master or test master is made and sent to you for your approval. You should listen to the audio in different places like your car, home stereo, and Walkman® to ensure the quality is good. If the recording is satisfactory, the recording or working master is sent back to the duplicator to make the number of copies ordered, and to package the tapes or CDs in plastic boxes.

Establish Time Frame and Completion Date for Project

You should establish a starting and ending date for your project. Make a list of the various contacts and bids you must make. For example:

- Audio script preparation 1-2 weeks
- Production/recording/editing 2 weeks
- Duplication 2-3 weeks
- Package design — Start at the beginning of the project. but allow two to three weeks.

Determine Your Budget for the Production

Example: For a 60 minute tape with pre-recorded music at the beginning, the end, and at the ends of the first side and the beginning of the second side, the sample budget below gives you an idea of expenses to expect. Actual costs could be more or less.

Voice script writer:	$200
Reader (author)	$000
Voice Coach	$125
Music	$300
Studio (record, edit, and mix down)	$800 to $1,000
Producer	$500 to $800
Tape materials	$200
Duplication, 500 tapes	$550
Graphic design	$300 to $400
Printing/cases	$800

Audio Album Package Design

Design is especially important if your audio cassette is to be sold in bookstores. Since a potential buyer cannot open up the cassette and read it, the design of the tape album/package itself is all there is to attract the person. An experienced designer should be hired who is familiar with the design of audio cassette albums. Some duplicators provide a design service. The package should be eye catching.

For marketing purposes it's a good idea to indicate on the package case that the tape is part of a series. If it is, a color pattern or series pattern should be used to indicate that more tapes are to follow.

Information to Put on the Front and Back Covers

Put the title and subtitle of the tape on the upper part of the front cover. Ensure that you acknowledge the name of the book or material that the information originally came from. The bottom part of the front cover should indicate the name of the publisher or producer of the tape.

On the back cover, the title should be repeated at the top and under that a summary of what the tape is all about. It should be interesting enough to whet the reader's appetite. If there is space, put in a short statement about the author, and at the bottom, a bar code which permits it to be processed rapidly at the cash register. The bar code indicates the price and the ISBN number or International Standard Book Number (assigned by the publisher), of the cassette or CD. The name of the producer/publisher should be near the bottom in small print.

Summary of Text to Be Included on Package Case Cover

You can, of course, rearrange the order of the following information, but in general this information should be included. Starting from the top down of the front cover.

Front Cover

1. Name or Title
2. Subtitle
3. Author
4. Illustration
5. Put number of cassettes in package (can go on back)
6. Name of publisher or series

Spine — Same as a book

1. Name or title
2. Author
3. Publisher
4. Series
5. Logo

Back Cover — Same as a book

1. Title of Tape
2. Summary about book and author

3. Names of readers, producer, publisher
4. Running time and number of Cassettes
5. ISBN Number
6. Price is optional as bar code has it
7. Copyright symbols (© and "P" or phono recording)
8. Bar code

Inside Front Cover

1. The title again if desired
2. More information about the tape
3. More information about the readers, if several, or the music, composers, singers and editors, etc.
4. Include the statement "All Rights reserved. Unauthorized duplication is a Violation of Applicable Laws."
5. Acknowledgements
6. If from a book, a statement that it is abridged or unabridged from such and such a book written by the author.
7. Name, address, and telephone number of the publisher

Supplies for Production

Prepare for your processing and shipping needs. Select and order the correct size mailers for shipping your package. If you, as publisher, are to do the shrink wrapping, make certain that you have needed supplies. You'd be wise to let the duplicator do the shrink wrapping. It can be a messy process if anything should go wrong. And doesn't it always?

Getting Recording and Package Bids

Just as having a book or material printed, obtain written bids from several studios, printing, duplicating and packaging companies to get the best prices. Keep in mind that pre-production cost may be more than your actual production costs.

Anticipating Demand for Your Audiotape/CD

It is important to estimate your anticipated market or audience for your project before you make an investment. Will it be your regular clients or target group, educators, students, professional or business organizations or a new target group? For a first tape or CD it is difficult to gauge the potential sales unless you've had some experience in the field. Keep in mind that many target groups, for example, the school market, take many months to recognize your product and to make a purchase decision.

Because the production of audiotapes from books in print is relatively new, sales of tapes and CDs are not as big as for books. This is changing as people become more familiar with spoken-word cassettes.

After you have determined your market area, decide how you are to reach the target group. Will you mail flyers, telephone clients, sell through distributors and bookstores, advertise in journals or radio, or sell direct after giving lectures?

4

RECORDING YOUR AUDIO–TAPE/CD

PRODUCTION PHASE

In the production phase, your voice material is read and recorded along with music and sound effects (SFX). As stated before, it is better to record in a sound studio specializing in voice to ensure a quality recording. Only pay for the equipment and services used in a studio.

Studio Recording

The publisher/producer should make certain that the sound engineer understands the purpose and process of recording the spoken-word on tape. If Digital Audio Tape (DAT) tape is to be used, make certain that the equipment is at the studio and that DAT recorder and tape is on hand (most studios charge additional for DAT use). In this process audio signals are encoded in a series of numerical digits that are stored on audio tape.

Some studios record on DAT tape and then transfer the best recordings to quarter-inch tape to edit. Digital editing or electronic editing is faster, but expensive equipment is required to do it well. Other studios will make the original recording on one-inch television tape, edit it electronically and transfer the finished tape onto quarter-inch tape.

Remote Recording

When a remote recording is necessary—the author/reader is in another location—you as a publisher may need to hire a studio for recording near them. In some cases a temporary studio can be established in a person's home, for instance, the author's home. Newcomers to recording often feel more comfortable in their own homes. The environment in one's home should be examined to ensure that the room is satisfactory.

What is a Quality Recording?

A professional quality tape means the recorded sound is intelligible, undistorted, and free from elements distracting to the listener. A quality tape production should have the following characteristics: It should be recorded at proper sound levels, not contain hiss or noises, and have no electronically induced sounds such as line hum or cross talk. The acoustic and ambient noises of the room where the voice was recorded should also be heard. The voice should have a presence.

One of the best ways to ensure a professional sound recording is to use experienced technicians and personnel from a production house or recording studio specializing in spoken-word or voice-over production.

Recording Process

Your material is recorded and edited on what is called an Original Master Tape. A copy (or copies) of this is made on what is called a Working Master Tape. The Original Master is put aside for safe-keeping. As an audiotape publisher you will need the Original Master tape and a Working Master. The Working Master tape is sent to the duplicator to make copies for inserting in cassettes or CDs.

How Voice, Music and Sound Effects Are Recorded

Usually, in producing a simple spoken-word production that contains only one reader, a producer can record everything on two-track tape. But if music and sound effects are added or many voices, then a multiple-track recording is necessary, requiring at least an eight-track recorder. Some producers need recorders with 36 tracks or more.

On multiple-track recording, usually a SYMPTE Time Code must be recorded first, then the voice tracks, then the music, and lastly the sound effects. This latter process is time consuming and can increase the cost many times. It has advantages because music can create mood and atmosphere. Sound effects give a sense of presence ("you are there") to the listener.

If you need to add "bell turns" for children's "read-along books," then plan for adequate time spacing and an open track for recording the bell turns. It could cost you valuable studio time to redo the spacing of voice tracks to accommodate bell turns. See the example below:

Simple Recording With One Voice

TWO-TRACK—This is usually produced in stereo (left and right)

- TRACK 1 - Voice
- TRACK 2 - Voice

Recording with Voice, Music, and Sound Effects

EIGHT-TRACK - Dramatized Book

- TRACK 1 -Voice 1
- TRACK 2 - Voice 2
- TRACK 3 - Voice 3
- TRACK 4 - Music
- TRACK 5 - Music
- TRACK 6 - SPX (Sound Effects)
- TRACK 7 - Blank (Buffer Track)
- TRACK 8 - SYMPTE Time Code

When a multiple track recording is mastered, the eight tracks (or more) are mixed down (or assembled) to two tracks and then transferred or dubbed to either another two track audiotape or DAT tape. This final material becomes the Original Master.

Doing a CD Recording

The same recording process and procedures are followed to produce an audio track for a CD-ROM as for an audio book or a CD-I for multimedia publishing. In CD-I multimedia publishing, the Original Master is the audio track to which text, graphics, and moving images are sequenced.

Understanding the order of producing your audio tracks can save you a lot of production headaches and money.

Use of Sound Libraries

As stated, above sound libraries are a source of material for your audiotape or CD. Libraries charge different rates for the different kinds of media uses, broadcast, non-broadcast, media presentations, and media for sale (for a spoken-word audio tape). It is better to pay a one-time fee for the use of music and not worry about the accounting of annual royalties. If music is in the public domain, you may not have to pay for permissions. Please check with a specialist or lawyer. The option of creating your own music may be the cheapest in the long run. A clearing house can help you to obtain rights for the use of music.

Tape Types

Have you ever wondered what bias tape to buy for your home stereo? High bias tape is used when music is to be recorded. For voice only, normal bias tape is satisfactory. But high bias is suggested for voice for better clarity. It costs a few cents more per cassette, but many producers believe it is worth it.

Studio Costs

Studio rental charges range from $15 to $200 or more per hour. The average voice-over studios charge about $35 per hour. The cost of a producer runs from $25 to $75 per hour. Most will work for a set fee. Studio sound engineers generally charge from $15 to $50 or more per hour depending on the studio and their facilities. The cost of the studio engineer is usually included in a sound studio's bill.

Production Considerations

Coordination

Many publishers make the mistake of using a college radio station or a music studio, with people not accustomed to working with a script and editing it. It takes experience and discipline for an editor to know when to splice a tape.

For every hour to be recorded, it may take two to three hours to edit. Many engineers and producers who do not specialize in this work tire easily and do sloppy work. A professional will maintain quality through the hours of the editing process and know how to pace it.

The process of selecting an audio production house can be facilitated by learning what to look for and what questions to ask. You should not try to be too thrifty in this area. Bad productions can jeopardize your product integrity and, for some, a line of products. Publishers new to the audio production process tell many horror stories about mistakes and going back to a studio to redo a reading.

If possible, use a specialist familiar with the voice-over format and type of program being created. Ask for the producer's experience and the type of equipment the studio has available. Ask what type of recording and editing equipment, and sound libraries they have.

Role of the Audio Producer

The producer should be well informed at all levels and experienced in audio producing, script preparation, recording, editing and post production. In particular, the producer's responsibilities are: 1. To review the project script for flow, grammar, and continuity; 2. To organize and direct the recording by monitoring the reader's performance and giving feedback on it; 3. To supervise the sound editing, addition of music and/or sound effects and the audio mix of voice and music; 4. If required by the publisher, to coordinate the cassette package design and printing, duplicating and assembling.

Sometimes producers are ill-prepared because the talent and/or the scripts are not in a comprehensible form. The scripts should be marked with production notes. Adequate and legible copies should be made available to all the people involved in production. Scripts should be typed, double spaced on 8 1/2" by 11" sheets of paper. Four to five copies of the script should be available to recording and studio personnel. A good engineer logs notes on the script as it is being recorded. Preferably, the engineer recording the production should be the one editing the production. Experienced producers know how to adapt to new situations and solve production problems.

Audio Script Format

There are two basic types of spoken-word audio tapes: abridged and unabridged, one with sound effects and the other without.

Who Provides the Speaking Voice?

Speaking voices can be provided by the author or publisher, or you can hire a media production company or an ad agency.

Voice and Image Releases from Talent

Either the publisher or producer should be responsible or request that the media house take care of obtaining releases. It depends on whether the publisher is in the same geographic area as the actors and the studio or miles away. If actors are hired, the publisher or media house should pay their fees. It is better to negotiate a fixed fee rather than provide royalties or residuals to the actors.

CHECK LIST

Recording, Cast, Music, and Sound Effects

Some of the questions to ask yourself before starting the recording or production phase:

1. Is music to be used?

If so, do you want:

_____existing music (sound libraries, original recordings)?

_____custom music?

2. Type of music style?

3. Number of voice characters in the production:______

Ages________,

Sex_________,

Voice qualities________?

4. Who will provide speaking voices?

______Author(s)

______Publisher

______Producer/sound studio

5. Who will secure voice and image releases from talent?

______Publisher or producer?

6. Who will secure actor agent and union agreements, if necessary?

7. Will a remote recording be necessary?

If so, where?

NOTES

5

POST-PRODUCTION

Editing and Sequencing

Post-production is the editing of the recorded audio voice and the mixing together of the voices, music, and sound effects (SFX). The mixing takes longer than the actual recording time. This is one of most important steps in the recording process. The sound engineers will edit for the ear and pace the talk. They will add in spaces, clean up the mouth sounds that people make when speaking, like smacking their lips, popping sounds, clearing their throat, etc.

If sound effects and/or music are used, the sound engineer must bring in the effects and music at the correct point in the recording and insert them or take them out at the correct time. When computers are used, producers and engineers "stripe the tape" (add a computer signal), and lay a SYMPTE time code before voice, music, and sound effects (SFX) are put on the tape.

Needle Drop Music Costs

Needle drop music can cost from $50 to $200 or more per drop. This can be for a few seconds or an entire song.

Digital Master DAT - Digital Audio Tape

Some producers/publishers like the completed Master on DAT (Digital Master Tape) because there is no signal loss in the duplication process. A blank DAT 120-minute tape costs about $15 to $20 each for recording. These must be copied in "Real Time." A DAT backup tape for duplication will be as good as the original.

Use of Dolby® Sound System

The Dolby® process eliminates the "hiss sound" from recorded tape. There are Dolby B and C. To use the process, a publisher must register with the Dolby company. This involves completing application forms, sending several copies of the tape, and details of the duplicator. A publisher is required to give credit to Dolby on the tape and to add the Dolby logo on the tape package and the cassette itself. The Dolby company will check products to ensure that publishers follow their rules.

6

PACKAGING AND MARKETING

Components of Tape Cassette Package

A tape package normally consists of the audio tape, the cassette containing it, a plastic box with a paper information card (called a "J" card), and a plastic or cardboard case, like a book cover, enclosing the cassette box. The CD is protected by a plastic cover that has a flat square shape.

Album Box Or Case

The audio album can be made from cardboard, plastic, or be vinyl molded. Many are customized. Libraries prefer a sturdy product. Book stores and jobbers like "hang tabs" on the albums for display purposes.

When you make plans for your audio album cover, make certain it is designed and printed as one piece to serve as the printing for both the outside and the inside of the front cover. When the printer inserts or glues the cover on the box, that one sheet will cover the outside and the inside of the box at the same time.

Either the manufacturer of the album or the printer will attach the cover printing to the album box. When seeking bids from a box printer or manufacturer for album boxes and printing, ask for samples. Also inquire whether they can make the box, print the cover material, attach the printing to it, assemble the package, and shrink wrap it.

A less expensive cassette container is made out of plastic but is not as durable as the library binding type. Instead of gluing information on the front of the box, the printed information, done on both sides, is inserted inside the plastic case. To save money, a publisher can, in-house, insert the printed cover inside the plastic case.

Plastic Box Holding Cassette

A group of cassettes can be placed in an album or "butterbox," as they are called, a style created by the HighBridge Company.

The cassette is placed in a hard plastic folding box called a Norelco box to protect it. It can be clear, partially clear, or black or various colors. Some soft plastic boxes, called "poly boxes," are used to hold the cassettes. We recommend the black, white or clear shells. Color is normally used for children, music, and other special tapes.

"J" Card

The "J" Card is a paper insert and is called that because from the side it's shaped like a J. It contains information about the tape inside the Norelco box. The "J" Card can be inserted inside the Norelco box, but it is not a necessity. If the needed information is put on the inside or outside of the plastic album case, the "J" Card can be eliminated and save you a lot of money if a large volume of cassettes is printed.

Information to Be Put Directly on the Cassette

You can print a label, with a sticky back, with needed information and attach it to the cassette, or you can have the information rubber stamped directly onto the cassette. The information should include the title of the tape, the author, the reader, the publisher(and address), copyright holder (include the year) of the audio, number of the side, and a statement indicating that "All rights are reserved," and that "Unauthorized reproduction is prohibited by law."

The duplicator company will normally put the information on the cassette after it has been loaded with recorded tape. Have the printer/duplicator send a sample of the design before imprinting. Once the label is imprinted it can't be erased. It would be very expensive to do another 1,000 or more over again. The rubber stamping process is called Apex imprinting. The publisher tells the duplicator what to put on the cassette. The publisher can put the sticky labels on the cassettes in-house if he wants to save a few pennies per cassette.

We recommend stamping your cassette shell because it is cheaper than using labels and has become the standard in the industry. Check that the tape is of good quality before stamping the shell. Also check that the printing is on the shell correctly and can be read. See the enclosed pricing list on page 47 for more details.

You should stamp the copyright © and the production "P" on each side of the cassette. The © covers the voice, script of book material and the "P" covers the "production" values such as the music, sound effects and recording procedures.

Mono or Stereo

You will also need to make a decision about whether to produce the tape in mono or stereo. If music or sound effects are used, stereo will certainly enhance the listening quality that the public seems to want. We recommend stereo as it provides a naturalness to voices.

Length of Audio Tapes

Tapes come in different time lengths. Generally they come in lengths of fifteen minutes. For example, C15, equals seven and one-half minutes on each side, C30, equals fifteen minutes on each side, C45, approximately twenty-two minutes on each side, C60, thirty minutes on each side, and C90, forty-five minutes on each side.

Quality of Tape — Premium or Normal

Depending on what the tape is to be used for, voice only, music, sound effects or a combination of sounds, different quality tapes are used. If music is used, you should consider using a chromium type tape. For voice only recording normal tape can be used. Since the use of high quality chromium tape costs only about three cents more per cassette, it may be wise to use quality tape even if it's for voice only.

Types of Cassette Shells

Cassette shells are usually of two types, the screw or the sonic-welded type. With the screw type, the two sides of the plastic cassette are held together by five screws. This is the desirable type because it holds up better with use over time.

With the heat welded type the plastic shell is pressured or welded together under a heat process. This type is cheaper, but the shell can sometimes affect the quality of the tape.

Transfers from Original Master to Working Master

After a Working Master tape has been made from the Original Master tape, the Original Master is put aside for safe keeping. The Working Master tape is sent to the duplicator who makes duplicating masters for producting the audio cassette or CD copies.

Tape Duplicating and Duplicators

To speed up the copying of tapes after a Working Master has been made, high speed duplicators are used by most tape duplicating companies. The duplicators copy information from the Master tape at high speed, about eight times normal speed, onto a large reel and then it loads the tape onto the audio cassette shells. The machine cuts the recorded tape and loads it into the cassette shells. The front and back of the tape are duplicated at the same time.

Some independent publishers have purchased high speed stereo duplicators to do their own duplicating for about $2,000 to $2,500. However, one problem with the cheaper high speed tape duplicators is that they can cause some dropouts, hiss, treble of voices, or make the pitch of voices go up. If you choose to buy a duplicator, you can duplicate tapes yourself based on the demand. But remember, this requires an investment in equipment and in time. If you elect to duplicate your own tapes, acquire a quality duplicator.

Real Time Duplication

To do a high quality tape, copies are made in "real time," which means at the same speed that the original tape was made. This process is more expensive, approximately three to four dollars per cassette versus about .60 cents per cassette for the high speed reproduction process. However, much better quality results from this process. Real time duplication is used when you have lots of music. Real time tapes are not needed for voice duplication.

Listen to Tapes from a Duplicator

Be aware that sometimes duplicators make mistakes and tapes sealed in cassettes may not be of good quality and have to be destroyed. It is important to at least spot check the quality of a batch of duplicated tapes if large numbers are being produced. In one case known to the authors, a tape run made of 20,000 cassettes had to be destroyed due to a mistake in duplication. Because the cassettes had been sealed, they couldn't even use the cassettes. It was a big loss to the duplicating company. In another case, the duplicator put the recordings on the tape in the wrong order such as side two on side one.

So Listen to your tape.

Remember to listen carefully to the Master tape in your own environment, car, Walkman®, home stereo, etc., and not just in a "quality" studio environment before duplication starts. Listen to the sound effects, voice, and music to ensure proper balance between them. In one case involving the co-author (producer), out of a run of 5,000 audio cassettes, one-third of them were dull and unusable due to a faulty duplicator. Most audio producers have horror stories to tell. Don't create any yourself.

Assembling the Tape Package

Assembling the completed audio tape and the case is a process that can take several weeks. Who will assemble the boxing of the tape in the package, the duplicator? If so, you must ship the box to him and pay him a fee to assemble the package, shrink wrap it, and ship it to you.

Shrink Wrapping

Shrink wrapping is important to protect the cassette from being used and put back in the package and returned to the store. Shrink wrapping enhances the shelf life of the product. If not wrapped, the package could look shopworn after a short time. People don't want to buy tapes or CDs that have been used.

If you choose, you can buy a shrink wrapping machine for about $400 and do your own shrink wrapping. A roll of plastic costs about $70 to $80. Two people can assemble about 100 tape packages and shrink wrap them in an hour, if nothing goes wrong. However, problems can occur. You can get pops in the plastic forcing you to take off the plastic and do it over again. It is best, if you can afford it, to pay the duplicator to do the shrink wrapping and avoid headaches.

Copyrighting Your Tape Cassette or CD

When you have completed your recording, apply for a copyright with the Copyright Office for a tape or CD cassette in the same way as you do for a book. Send two copies to the Copyright Office with the required payment (now $20). Make a decision as to who is to file and hold the copyright for the project. Call or write the Register of Copyrights, Library of Congress, Washington, DC 20559, for copyright form SR (sound recording).

Sample Cost of Audio Services and Equipment

Included below are sample costs from a duplicator and packaging company for audio duplication for different lengths, printing, and packaging. They are for orders of 1,000, 3,000, 5,000, and 10,000. (Information provided by courtesy of Multi Media, Publishing and Packaging, 19360 Rinaldi St., Suite 321, Northridge, CA 91326)

Services

- ❑ Professional Master Transfer (Standard ¼", DAT, PCM)
- ❑ J-Card, Label Design & Typesetting
- ❑ Typesetting for direct on-cassette printing
- ❑ Color Separation and Color Printing
- ❑ Premium Normal & High Bias Tape
- ❑ Choice of Cassette shells and Norelcos
- ❑ Customized Packaging & Shrinkwrapping

Pricing Worksheet

A. Audio Duplication

	1000	3000	5000	10000
C-15	.36	.35	.34	.33
C-30	.40	.39	.38	.37
C-45	.45	.44	.43	.42
C-60	.49	.48	.47	.46

Included in above Pricing:

- ❑ Duplication using Premium Normal Bias Tape
- ❑ Choice of Five Screw or Sonic Welded Cassette Shell
- ❑ Transfers from Original Master to Working Master

B. Printing and Packaging

	1000	3000	5000	10000
J-Card	.062	.056	.052	.050
A+B Labels	.031	.027	.025	.021
Apex Print	.025	.022	.018	.016
Norelco Box	.13	.12	.11	.106
Polybox	.087	.08	.065	.075
Shrinkwrapping	.018	.016	.012	.010

MARKETING

Prepare a plan for marketing your tape or CD as you would for a book or any other product. With an audiotape or CD you use a slightly different approach. One of the first things to do is to list who the audio tape reviewers are for national reviewing publications such as Publishers Weekly, Library Journal, etc. These two publications are widely read by librarians to learn of new audio products. Do the same for the wholesalers and distributors such as Baker and Taylor, Ingram, Inland, Publishers Group West, Bookpeople, Pacific Pipeline, etc.

Most audio book cassettes are sold through direct mail. If you want to be included in direct mail catalogs, send promotional material four months in advance of the publication date. Also become aware of audio book listing deadlines at Publishers Weekly and other publications.

Review your national and regional mail list and select from it those persons or groups that might be interested in the subject of your tape. Develop your own special market mail list and hit it with announcements, fliers, or any promotional methods that have worked for you. Once funded, implement your plan.

To enhance your publishing background in audiotape join organizations like PMA (Publishers Marketing Association). They publish a monthly newsletter that is read by over 4,000 publishers, writers and distributors. The members are very friendly, network a lot, and can give you helpful advice. Members are spread throughout the United States. Contact Jan Nathan at (213) 372-2732.

Another organization to consider contacting is Audio Publishers' Association located in New York.

Some Marketing Suggestions

Printed Word Advertising

You should mention on your tape or the tape package something about the book or material that the recording came from. To market your tape, use a similar approach to selling a book. The only difference is that a publisher will promote it with different people and write up promotional material a little differently.

Obtaining an Audio Tape Review

As for a printed book, printed seminar program, or your special printed material, send your audio cassette in advance of official publication date to audiotape reviewers at Publishers Weekly, Library Journal, or special interest journals, etc., to obtain reviews. Consider sending announcements to reviewers before you actually send the tape itself. Include a "bounce back" reply card for a free or demo copy of your audiotape that includes an order form or business reply form.

Demo Tape

Because reviewers may not want to listen to an entire tape, you may want to send them a short demonstration tape. Select the sections you want and ask your editor to splice it together.

Distributors

When contacting a distributor, send the complete album package rather than just the audiotape. They want to see what kind of album the book stores, libraries and other markets will receive. Also they are not as apt to listen to a full tape before making a decision to handle your project. The distributor is concerned about the attractiveness and durability of the package to the customer. No one wants returns. Tell your distributor that you won't accept customer opened cassettes unless the cassette is defective.

See Appendix, Resources section, page 80, for list of distributors that handle audio cassettes.

Flyer and Bounce Back Cards to Target Groups

An economical and effective way to reach your target group is to send them a well-designed flyer with an order form on it or a bounce back card with an order form. Unless you know your book stores, it is probably best to sell to distributors rather than to sell direct. Keep in contact with your distributors about your audiotapes.

Tape Sales

The bulk of audiotape sales are through direct mail. Make an effort to reach your present customers who bought printed books or material from you. A potential market is with audio/video rental stores. Try marketing to video, children's book and clothing stores. Another possibility is gift shops and hospital gift shops for tapes concerning health, injuries, grief, and spiritual matters. Truck stops are another possibility. Truck drivers buy large quantities of tapes and it is not all country music. With more women driving big rigs, they are buying many spoken-word type tapes.

You can always rent or buy a specialized mailing list. Try the libraries, colleges and secondary schools, and professional organizations. Don't forget to send promotional material to your competitors if appropriate.

Many publishers release and market their book and audio tapes at the same time. Both products can benefit from one promotion effort. For the first run of a tape, it might be wise to limit it to 500 copies or less to test the market. Short runs of audiotapes are not as expensive as it is for book printing. Remember to mention something about the book on the audio tape package.

Sales Price of Your Audio Cassette or CD

At present, the sales prices are running about the same, $9.95, for a one-cassette package of 90 minutes, $14.95 to $19.95 for a two-cassette package of approximately 180 minutes, and up according to the number of cassettes. This pattern can change, of course. If possible, try to keep your tape project down to a two-cassette package and away from a five or six-cassette one. The public seems more interested in paying for a one or two-tape package although sales have been steady for three or more cassette albums. Whether a tape has music or sound effects does not normally justify increasing the cost above the norm unless a particular tape is unique and in demand by people willing to pay for the information. Schools and libraries will sometimes pay more than the norm for tapes they want.

CHECK LIST

Tape Package

The publisher, author, or media production house must decide on the following:

- Audio cassette/CD packaging needs
- Package artwork, graphic design
- Will the cassette need a sticker label or apex imprint printed on the cassette__________?
- No. of colors_____
- Norelco box_____
- Clear_____
- Clear/black_____
- Plastic Box_____
- Color of cassette shell_____ Clear_____

Duplication

- Who will provide the duplication facilities? Publisher or media production house
- What type of cassette tape will be used?
- Low noise cassette tape____ Chrome cassette_____
- Length of tape—C-15, C-30, C-45, C-60, C-90

High Speed or Real Time duplication (A production with extensive musical production should be duplicated in Real Time. Most voice only productions can be duplicated at high speed).

- Duplication using premium or normal bias tape?
- Five-screw or sonic welded cassette shell?
- Transfers from original master to working master?
- Standard one-quarter inch or DAT recording tape?
- Dolby or non-Dolby noise process?

NOTES

7

CONCLUSION

If the first year of the 1990s is a forecast of the rest of the 1990s, the spoken-word audiotape field will be an exciting one in this decade. Spoken-word audio sales have been strong and showing signs of expanding into new markets. Sales moved beyond the bookstores into airport stores, truck stops, supermarket chains, record stores, and video outlets.

New audio-only stores evolved. The Jimmy B.'s Audio Books started in two locations in California. In Chicago, Words in Motion opened in 1990 and the owner planned new outlets.

In Publishers Weekly's January 4, 1991 issue, entitled *1990: A Review,* several spoken-word audio companies were interviewed for their views on 1990 and the trends for the future. Michael Viner, President of Dove, said that 1990 was "the best year ever" for audio sales. The Harper-Audio Company announced that they had entered into an agreement with a major restaurant chain to sell their audio products. George Hodgkins of Audio Renaissance Tapes stated, "Libraries definitely started to show more interest this past year." Jim Brannigan, Director of Marketing for HighBridge Productions said, "Every year we say we're poised for growth, but now other markets are noticing us. We're seeing interest from the record stores and video outlets." Others in the article voiced similar opinions concerning the growth of the audio industry.

Traditional book buying customers are not only buying more audiobooks, but are also requesting specific audio book authors and titles from bookstores. Oxford Books in Atlanta indicated that "Customers seem to be more aware of audio in general. . ." Booksellers found that fiction continued to dominate the audio market in 1990, but self-help and business titles were strong. For example, Dove's *A Brief History of Time* by Stephen W. Hawking sold 100,000 tape books in 1990. Other books such as *All I Really Need to Know I Learned in Kindergarten* and *It Was on Fire When I Lay Down on It* also hit the 100,000 mark.

Trends

New audio categories appear to be opening up. Autobiographys, such as Gilda Radner's, *It's Always Something*, true crime, and unabridged works are gaining momentum. In general, people in the audio field are convinced that audio is headed for even greater expansion as it gains greater and greater acceptance by the book-buying public.

Because of greater competition within the industry from the well established publishers, and the innovative and creative minded independent publishers, anyone entering the field must think quality in the production of their creative works. By being able to find answers to the questions presented in this book, you should be able to set goals for reaching your target population, prepare a budget, production schedule, and market plan. You should also know about the terminology and procedures. Of course there is always much more to learn, but you are ready to move ahead in this exciting field. **GOOD LUCK!**

APPENDIX

GLOSSARY

AA - "Acting Ability"

ANALOG - Traditional recording on magnetic tape based on frequency soundwaves (as opposed to the numerical digital recording).

AUDIOBOOK - Spoken-word version of printed text book.

AUDIONOVEL - Spoken-word version of printed text novel.

AUDIO TAPE PACKAGE - Includes tape, cassette, Norelco box, and album case.

APEX PRINTING - Term from the APEX Company which imprints text directly onto audio cassette shells.

ARRANGER - Arranges the instrumentation (nowadays synthesizer samples and sequencers).

BETAMAX - Sony one-hour, 1/2" videocassette recorder.

BIAS - A steady electrical voltage—A.C. or D.C.—applied to an electronic circuit to make it operate more efficiently.

BOOKS ON TAPE - Trademark name for audio publishing company in California. Do not use term in name or marketing material.

BOOM - A microphone stand with a long arm that suspends a microphone over the action.

BUTTERBOX - A term for a series of audio cassettes packaged together (coined by Jim Brannigan at HighBridge).

CADENCE - The breaks between words and the phrasing of words aligned together.

CD - Compact Disc - This is an audio format that is recorded on 12 cm optical discs which offer high fidelity sound.

CD-ROM - Compact Disc - Read Only Memory.

CD-ROM XA - CD-ROM Extended Architecture. A recording format that allows digital audio to be compressed at lower qualities allowing the capacity to increase storage of digital information from four to nineteen hours on a single CD.

COPYRIGHT - The right of an author of a creative work to enjoy the rewards of his efforts, in particular the right of a musician to the proceeds of the work he has composed or recorded as defined by the Copyright Act. Registration of material with the Library of Congress. For audio recordings use FORM SR (Sound Recording).

DAT - Digital Audio Tape. Audio signals are encoded in a series of numerical digits that is stored on audio tape.

DIGITAL EDITING - The process of editing through numerical digits (as opposed to ANALOG or traditional razor blade editing).

DROPOUT - A loss of audio and/or video signal during tape playback.

DUB - Tape to tape transfer.

DUBBING - Duplication of the audio tape to another tape or the addition of new audio material to the tape.

FADE - A gradual change in the sound volume level (increase or decrease).

FINAL MIXED STAGE - All items, voice, music and sound effects have been mixed on to one tape.

EDITING - Technique of improving original recordings by the removal of unwanted sections, the insertion of new material, and the introduction of special effects.

HIGH SPEED DUPLICATION - Duplication of audio cassettes at high speed.

ISBN - International Standard Book Number.

MAGNETIC TAPE - The sensitive material on which the tape recording is made. It is coated with various materials such as magnetic iron oxide and other materials to contain the recording.

NORELCO BOX - Hard clear plastic box that stores a single audio cassette.

NEEDLE DROP MUSIC - Refers to the per use of prerecorded licensed music and sound effects.

ORIGINAL WORKING MASTER - Tape on which the original or completed final recording is located.

WORKING MASTER TAPE - A copy of the Original Working Master. Used in making copies.

POLY BOX - Soft plastic box that stores a single audio cassette.

PFM - Per Finished Minute

REAL TIME DUPLICATION - Duplication done at actual time (i.e., if a tape is 30 minutes in duration it will take approximately 30 minutes to duplicate).

RELEASES - Permission and clearance forms. See Appendix.

VISUAL RELEASE - A release obtained from artist who prepared visuals for a production.

VOICE RELEASE - A release obtained from person or actor who provided voice for production.

SYNCHRONIZATION RELEASE - Release form for published recorded music with visual images.

COMPOSITION/WORK FOR HIRE RELEASE - Release form for "buy out" of musical composition.

SOUND - Audible vibrations, usually of the air, set up by any form of rapid cyclic movement having a frequency lying between 20 and 20,000 vibrations per second.

REMOTE RECORDING - Recording voice or sounds outside of the studio. For spoken word this could be in a quiet home, church, or any facility that is soundproof. Not advisable unless you know what you're doing.

SESSION - Term for a recording date.

SHELL - Another term for the audio cassette enclosure.

SHRINK WRAP - Clear plastic that is wrapped around products. By using high temperature, either hand-held dryer or a heat tunnel, plastic shrinks to form the package and gives it a glossy glow. It also protects the product.

SPOKEN WORD - Industry term for books-on-cassettes or CD ROM (also audio books, audio novels).

STEREO RECORDING - Two or more channel recordings of audio sound. For audio cassettes, usually left and right channels. Gives a more realistic sound.

SCRIPT EDITOR - Person that writes and edits audio script for the ear.

SFX - Sound Effects. Use as a notation on audio and video scripts.

SIGNAL - Information registered as an electrical impulse.

SOUND LIBRARY - Prerecorded licensed music and sound effects available for use.

SOUND POSTURING - A phrase used by poet Robert Frost that refers to meanings conveyed by sounds.

SYNTHESIZER - Electronic instrument producing musical sounds.

STRIPING THE TAPE - The process by which SYMPTE time code is recorded onto one of the audio tracks. This must be done first as a reference for voice, music, and sound effects..

SYMPTE - The Society of Motion Picture and Television Engineers that sets standards for the industry.

SYMPTE TIME CODE - A code that uses 8 digits for individual frame identification of the form: HH:MM:SS:FF (hours, minutes, seconds, frame numbers).

TRADE PUBLISHERS - Book publishers selling to book stores and other groups.

TIME CODE - The SYMPTE code recorded on an audio track that allows the computer to locate a specific frame address.

TRACK - The actual area on a magnetic tape that carries the recording. Can range from one or two to eight or more.

VOICE COACH - A speech specialist that works with reader and producer to coach and ensure quality control on the clarity of the spoken word.

VOICE OVER (VO) - A spoken message or narrative.

VCR - Video Cassette Recorder

VHF - Very High Frequency (television transmission signals).

SAMPLE AUDIO PRODUCTION AGREEMENTS

Please note that the following author, reader, actor, musician, and publisher/author agreements, work for hire agreements, and release forms, are only suggestions for the reader to consider. Each audio producer/publisher should seek legal counsel on what to include in any agreements.

AUTHOR CONTRACT AGREEMENT BETWEEN

____

and

____

AN AGREEMENT made between ____________________________(Author), a citizen of the United States of America whose home is anywhere USA, and ____Publisher of____ ________________________on January 1, 1991.

THE AUTHOR AND PUBLISHER AGREE THAT

1. The Author will write for publication a work to be named ________________________ in an audio script format. The Author grants this work to the Publisher with the exclusive right to publish and sell the work, under its own name and under other imprints or trade names during the full term of copyright and all renewals thereof, and to copyright it in the Author's name in all countries; also the exclusive rights listed in paragraph I below; with exclusive authority to dispose of said rights in all countries and in all languages.

2. The manuscripts for the audio script will be delivered to the Publisher by the Author by _________, 1991.

3. When the manuscript is ready for publication, it will be produced and published at the Publisher's own expense. The Publisher will pay the Author a royalty based on the actual gross receipts cash received by the Publisher of ___%.

4. The Publisher will report on the sale of the work on the first of March and September of each year for the six-month period ending the prior December 31 and June 30, respectively. With each report of sales, the Publisher will make settlement for any balance shown to be due.

5. Paragraphs A-R inclusive, on pages following, are parts of this agreement as though placed before the signatures.

____

John Doe, Author, Date______

____

Social Security Number

____

Publisher, Date ____________

A. The Author will deliver the manuscripts in typewritten form and on computer diskette. The manuscripts will be submitted in duplicate and a third copy will be retained by the Author. It will be in proper form for use as a copy by the printer, and the content will be such as the Author and Publisher are willing to have appear in print. If the Author fails to deliver a satisfactory manuscript on time, the Publisher will have the right to terminate this agreement and to recover from the Author any sums advanced in connection with the work. Until this agreement has been terminated and until such sums have been repaid, the Author may not have the work published elsewhere. The Author will read the proofs, correct them in duplicate, and promptly return them to the Publisher. The Author will be responsible for the completeness and accuracy of such corrections and will bear all costs of alterations in the proofs (other than those resulting from printer's errors) exceeding 10% of the cost of typesetting. These costs will be deducted from the royalty payments due the Author.

B. The Publisher will furnish the following items: front and back cover copy on the packaging of the audio tape; preface or forward (if any); index; and complete and final copy for all illustrations properly prepared for reproduction.

C. The Author warrants that the Author is sole owner of the Work and has full power and authority to copyright it and to make this agreement; that the work does not infringe any copyright, violate any property rights, or contain any scandalous, libelous, or unlawful matter. The Author will defend, indemnify, and hold harmless the Publisher against all claims, suits, costs, damages, and expenses that the Publisher may sustain by reason of any scandalous, libelous, or unlawful matter contained or alleged to be contained in the work, or any infringement or violation by the work of any copyright or property right; and until such claim of suit has been resettled or withdrawn, the Publisher may withhold any sums due the Author under this agreement.

D. The work will contain no material from other copyrighted works without the Publisher's consent and the written consent of the owner of such copyrighted material. The Author will obtain and be financially liable for such consents and file them with the Publisher.

E. The Publisher will have the right to edit the work for the original printing and for any reprinting, provided that the meaning of the text is not materially altered.

F. The Publisher will have the right: (1) to publish the work in a suitable style as to paper, printing, and binding; (2) to fix or alter the title and price; (3) to use all customary means to market the work.

G. The Publisher will furnish ten (10) copies each of the completed audio tape to the Author without charge. Additional copies for the Author's use shall be supplied at a _____ percent (__%) discount from the lowest list price.

H. The Author agrees to revise the work if the publisher considers it necessary in the best interests of the work. The provisions of this agreement shall apply to each revision of the work by the Author as though that revision were the work being published for the first time under this agreement. Should the Author be unable or unwilling to provide a revision within a reasonable time after the Publisher has requested it, or should the Author be deceased, the Publisher may have the revision prepared and charge the cost, including, without limitation, fees or royalties, against the Author's royalties, and may display in the revised work and in advertising the name of the person, or persons who revise the work.

I. The Publisher may permit others to publish, broadcast by radio, make recordings or mechanical renditions, publish book club and microfilm editions, make translations and other versions, show by motion pictures and by television, syndicate, quote, and otherwise utilize this work, and material based on this work. The net amount of any compensation received from such use shall be divided equally between the Publisher and the Author.

The Publisher may authorize such use by others without compensation, if, in the Publisher's judgment, such use may benefit the sale of the work. If the Publisher itself uses the work for any of the foregoing purposes (other than publishing), the Author will be paid __% of the cash received from such use. On copies of the work or sheets sold outside the United States, the Author will be paid a royalty of __% of the cash received from such sales. On copies of the work sold through any of the Publisher's book club divisions or institutes, or by radio, television, mail-order, or coupon advertising direct to the consumer or through any of its subsidiaries, the Publisher shall pay the Author a royalty of __% of the cash received from such sales. If the Publisher sells any stock of the work at a price below the manufacturing costs of the audiotape plus royalties, no royalties shall be paid. All copies of the work sold and all compensation from sales of the work under this paragraph shall be excluded in computing the royalties payable under paragraph 3 above and shall be computed and shown separately in reports to the Author.

J. If the balance due the Author for any settlement period is less than ten dollars, the Publisher will make no accounting or payment until the next settlement period at the end of which the cumulative balance has reached ten dollars. If, after the expiration of two years following publication, the sales in any twelve month period ending December 31 do not exceed a total of 500 copies, the royalty on such sales shall be one-half the stipulated royalty being agreed upon to enable the Publisher to keep the work in print as long as possible. The Publisher may deduct from any funds due the Author, under this or any other agreement between the Author and the Publisher, any sum that the Author may owe the Publisher. When the Publisher decides that the public demand for this work no longer warrants its continued manufacture, the Publisher may discontinue manufacture and destroy any of all audio cassette tapes and master tapes without liability to the Author.

K. The Author agrees that during the term of this agreement the Author will not agree to publish or furnish to any other publisher any work on the same subject that will conflict in the sale of this work.

L. This agreement may not be changed unless the parties to it agree in writing.

M. This agreement shall be construed and interpreted according to the laws of the State of ________________ and shall be binding upon the parties hereto, their heirs, successors, assigns, and personal representatives; and reference to the Author and to the Publisher shall include their heirs, successors, assigns, and personal representatives.

N. The Publisher agrees to make no initial payment to the Author as an advance against royalties will be payable as follows: upon the signing of this agreement and upon the acceptance by the Publisher of the completed manuscripts in a form, style, and content acceptable for publication; provided, however, the Publisher may retain for its investors the payments due until paid in full otherwise due the Author under the terms of this agreement. The terms of this paragraph shall not apply to revised editions of the Work.

O. If completed manuscripts, acceptable to the Publisher, shall not have been delivered by_______, 1991 or within an additional ninety-day grace period, by reason of the Author's death or otherwise, the Publisher may terminate this agreement, whereupon all monies paid to the Author shall then be repaid to the Publisher. The terms of this paragraph shall not apply to revised editions of the work.

P. If the completed manuscripts have been accepted by the Publisher and there is no evidence of any progress to publish the work within six months upon signing of this agreement, the Author may terminate this agreement and the Publisher agrees to revert all rights back to the Author.

Q. Upon written request of the Author after the audiotape has been declared out of print, the Publisher agrees to revert all rights back to the Author. It is generally understood that "audio rights" are granted for two (2) year terms (from the date of publication) and renews itself automatically unless either Author or Publisher agrees in writing to cease the relationship.

R. A photocopy of this agreement shall have the same force and effect as the original.

MEDIA PRODUCTION SERVICE AGREEMENT

AGREEMENT BETWEEN

_______________________________________Press

AND

____

(hereinafter "X"), a ______________Corporation with principal place of business located at ______________________________, and of ________________________(hereinafter " "), as Independent Contractor with principal place of business located at____

WHEREAS, X intends to contract with _________ for the performance of certain media production services.

WHEREAS, _____________represents that it is engaged in an independent calling and has complied with all federal, state, or local laws regarding business permits, self-employment taxes, insurances and licenses that may be required to carry out said independent calling and the tasks to be performed under this agreement;

WHEREAS, _______ represents that it is engaged, or will be in the foreseeable future, in the same or similar activities for other clients and that X is not _______'s sole and only client or customer.

THEREFORE, IN CONSIDERATION OF THE FOREGOING REPRESENTATIONS AND THE FOLLOWING TERMS AND CONDITIONS, X AND _______ AGREE:

1) "X" engages _______ to perform media production services to complete the _ Spoken-Word Audio Cassettes.

2) "X" agrees to pay $ _______ for services rendered according to the fee schedule of ITEM A (Attached) and by an invoice which _______ will submit for material expenses and work performed.

______'s invoices for work performed will be due on the 1st of each month. Invoices should reflect only materials expenses and work done in the previous month. Invoices will be processed and paid within 15 days. Services will be charged according to the fee schedule of ITEM A (Attached).

Any changes to "FINAL MIXED" media materials (see Paragraph 6) would be billed as additional to "X" as:

$ 55/Hour for Studio Time

$ 40/Hour for Media Producer

$ 25/Hour for Music Producer

and also including direct sub-contractor costs for services rendered and materials.

3) _______ shall generally coordinate its work with "X" keeping "X" informed of current and anticipated activities through weekly reports. _______ general activity shall be directed by "X" although the manner, specific schedule and mode of activity remain discretionary with _______, except as requested for scheduled meetings with "X" or scheduled times particularly for the joint performance of certain tasks. The parties agree that this time shall be served during the months of January and February 1991 for the term of this agreement, although both may either exceed or not achieve this time commitment.

4) "X" will provide _______ no administrative services, except for reimbursement for telephone, photocopies, and mailings.

5) _______ retains the sole right to control or direct the manner in which the services described herein are to be performed. Subject to the foregoing, "X" retains the right to inspect, to stop work with reasonable notice, to prescribe alterations and generally, to supervise the work to insure its conformity with that specified in this agreement.

6) WHEREAS, "X" retains the right to prescribe alterations in media formats prior to "FINAL MIXED" stage. It is

agreed that any changes made after "FINAL MIXED" stage that _______ shall submit invoices to "X" in accordance with Paragraph 2.

7) _______ and "X" understand that it is _______'s sole responsibility and not "X's" responsibility to provide security and insurance, including worker's compensation coverage for their employees and personal liability insurance arising out of or relating to this agreement.

The parties understand and agree that _______ shall secure and also maintain their own insurance; pay their own expenses (except for preapproved "X" objective) (see Paragraph 4); report their income and in all other respects generate independently and not as employee of "X".

8) "X" agrees to indemnify and hold harmless _______ from any claim by any 3rd party against "X" for any act or non-act during the life of this agreement. Under no circumstances shall _______, their successors or their assigns be held liable for claims, judgments, actual or consequential damages, or attorney's fees arising out of governmental or civil actions, including but not limited to those involving misrepresentation, omission and fraud related to services performed for "X."

9) _______ retains the right to refuse individual assignments without prejudice to the overall agreement in the event of circumstances such as conflict of interests, work overloads, lack of capabilities, etc.

10) _______ will submit his federal tax ID numbers to "X" at the time this contract is signed.

11) At the end of the calendar year a 1099 form will be sent to ___for all invoices paid in excess of $600 for that year.

12) No FICA (social security) payroll tax shall be withheld from the payments to _______ by "X."

13) No federal unemployment insurance shall be paid by "X" on account of_______.

14) No state or federal income tax shall be withheld from the payments to _______.

15) Neither State Disability Insurance (SDI) nor State Unemployment Insurance (SUI) shall be withheld or paid by "X" on account of _______.

16) No unemployment compensation will be granted by "X", if the need for services terminate.

17) No worker's compensation insurance has been or will be obtained by "X" on account of _______ for the tasks to be performed by _______ and their employees.

18) "X" shall pay _______ in accordance with Paragraph 2.

19) This agreement can be terminated by either party upon two weeks' written notice to the other, signed by the party electing to terminate.

20) There will be a severance fee of $ _______. In case of termination, a fee will be paid to _______ by "X" based on expenses for materials and services up to and including the date of termination.

21) Any dispute between _______ and "X" concerning the terms and conditions of this agreement shall be submitted to mediation, and, if necessary, binding arbitration pursuant to the laws of the State of _________.

22) Conditions stipulated in any verbal or written authorization shall not override the preceding terms unless they have been specifically agreed to in writing wholly separate and unique to the item in question.

23) This agreement shall begin on January 1, 1991 and continue until February 28, 1991.

24) Any activities beyond the scope of this agreement shall be the subject of a new agreement.

This agreement constitutes the entire and complete understanding of the parties hereunto. Any changes, modifications or alterations of the terms here in must be in writing and signed by each party.

A photocopy of this agreement shall have the same force and effect as the original.

"X" ______________________________ Date__________

By: ______________________________

XX ______________________________ Date __________

ACKNOWLEDGEMENT BY ARRANGER OF WORK-FOR-HIRE STATUS

Whereas ________________________ (hereinafter "arranger"), has at the request of ____________________ (hereinafter "_______")and upon _______'s special order and commission made the following musical arrangements:

Whereas, the arranger and _______ agree that the said arrangements shall be considered as works "made for hire" as contemplated and defined in Section 101 of the United States Copyright Act:

Now, therefore, in consideration of the sum of _____________ ($ dollars), and other good and valuable consideration, receipt of which is hereby acknowledged, arranger does hereby acknowledge such employment, and agree that under the terms of such employment such arrangements and all rights appertaining thereto are entirely the property of _______, its successors and assigns, absolutely and forever, for any and all copyright terms and all extension and renewal terms of copyright whether now known or hereafter created throughout the world, and for all uses and purposes whatsoever and free from the payment of any royalty or compensation whatsoever,and credit may be given for said arrangements to the Arranger in whole or in part, or not at all at the sole discretion of _______.

In witness whereof, the parties have hereunto agree to the above on this day ___________ of ______________________.

Signature of Arranger____________________________

Signature of ________________________ (Employer)

By ____

RELEASE (IMAGE OR VOICE)*

In consideration of receiving $ _____ for a recording session and other good and valuable consideration, receipt of which is hereby acknowledged, the undersigned hereby sells and assigns to ___________________________ its agents, affiliates, licensees, successors and assigned, the right and authority to use on radio, television, motion picture, audio tape or cassette, video tape or cassette, theatrically, in print, and in any other media, and in the promotion or exploitation thereof, reproductions of my image and/or voice made or produced between (DATE)__________________ and (DATE)__________________ at (LOCATION)_______________________________,and consent to the use of my name in connection therewith.

I hereby release _____________________________ its agents, affiliates, licensees, successors and assigns, from any and all claims and causes of action that I presently have or may have in the future relating to my participation in the production above mentioned, including but not limited to residual rights to which I may be otherwise entitled. ____________________________is free to exercise its artistic control in preparation of the final product or products derived from my work in the production. ____ may edit or alter the product, use portions thereof or none at all, all within their sole discretion.

All or any part of _____________________________ rights hereunder may be licensed, transferred or assigned by ____________________, its respective agents, affiliates, licensees, successors or assigns. I hereby agree to execute, acknowledge and deliver any and all additional assignments and other instruments which in the sole judgment of ______________________________may be deemed necessary or expedient to effectuate the purpose or intent of this agreement.

Interpretation of this agreement shall be according to the laws of the State of ________________. If any portion of this agreement shall be determined invalid or unenforceable, the remainder of the provision shall remain in full force and effect and shall in no way be affected, impaired or invalidated.

A photocopy of this authorization shall have the same force and effect as the original.

Signature ___________________________Dated: ———

____

Print Name

____

Social Security Number

____

Street

____

City, State, Zip Code

Phone (____)_________________________

* Provided by MEDIA BRIDGE TECHNOLOGIES

RELEASE (IMAGE OR VOICE) FOR CHILDREN/MINOR*

In consideration of receiving $ _____ for a recording session and other good and valuable consideration, receipt of which is hereby acknowledged, the undersigned hereby sells and assigns to ___________________________, Publisher, its agents, affiliates, licensees, successors and assigned, the right and authority to use on radio, television, motion picture, audio tape or cassette, video tape or cassette, theatrically, in print, and in any other media, and in the promotion or exploitation thereof, reproductions of my image and/or voice made or produced between (DATE)_________ and (DATE)____________________ at (LOCATION) ____, and consent to the use of my name in connection therewith.

I hereby release ____________________________, Publisher, its agents, affiliates, licensees, successors and assigns, from any and all claims and causes of action that I presently have or may have in the future relating to my participation in the production above mentioned, including but not limited to residual rights to which I may be otherwise entitled.

____________________________, Publisher is free to exercise its artistic control in preparation of the final product or products derived from my work in the production. ________________________, Publisher may edit or alter the product, use portions thereof or none at all, all within its sole discretion.

All or any part of ______________________, Publisher's rights hereunder may be licensed, transferred or assigned by ______________________________, Publisher, its respective agents, affiliates, licensees, successors or assigns. I hereby agree to execute, acknowledge and deliver any and all additional assignments and other instruments which in the sole judgment of ______________________, Publisher, may be deemed necessary or expedient to effectuate the purpose or intent of this agreement.

Interpretation of this agreement shall be according to the laws of the State of ________________. If any portion of this agreement shall be determined invalid or unenforceable, the remainder of the provision shall remain in full force and effect and shall in no way be affected, impaired or invalidated.

A photocopy of this authorization shall have the same force and effect as the original.

Dated: ________________________________

____

Minor's Signature

____

Print Name

____

Social Security Number

____

Home Address

____

City, State, Zip Code

GUARDIAN CONSENT

I, the undersigned, being the parent or guardian of ______________________________, do hereby consent to the grant of authority and signature thereto as shown on this form above.

______________________________(_____)________

(Signature)(Date)(Telephone Number)

____

(Address)(City, State, Zip Code)

* Provided by MEDIA BRIDGE TECHNOLOGIES

RELEASE - WORK FOR HIRE/ COMPOSITION/ ARRANGEMENT PERFORMANCE*

____________________________, Publisher

____

____

RELEASE FOR MUSICAL COMPOSITION/PERFORMANCE

In consideration of receiving the amount specified in Article A (attached) and other good and valuable consideration, receipt of which is hereby acknowledged, the undersigned hereby sells and assigns to __________________________, Publisher, its agents, affiliates, licensees, successors and assigned, the right and authority to use on radio, television, motion picture, audio tape or cassette, video tape or cassette, theatrically, in print, and in any other media, and in the promotion or exploitation thereof, reproductions of my compositions and performance as described as "synthesizer keyboard" during the month of February 1991 at the ___________Recording Studios in _______________ and consent to the use of my name in connection therewith.

I hereby release __________________________, Publisher, its agents, affiliates, licensees, successors and assigns, from any and all claims and causes of action that I presently have or may have in the future relating to my compositions in the production above mentioned, including but not limited to residual rights to which I may be otherwise entitled.

________________________________, Publisher, is free to exercise its artistic control in preparation of the final product or products derived from my work in the production.

______________________________, Publisher, may edit or alter the product, use portions thereof or none at all, all within its sole discretion.

All or any part of ____________________________, Publisher's rights hereunder may be licensed, transferred or assigned by ________________________, Publisher, its respective agents, affiliates, licensees, successors or assigns. I hereby agree to execute, acknowledge and deliver any and all additional assignments and other instruments which in the sole judgement of ____________________, Publisher, may be deemed necessary or expedient to effectuate the purpose or intent of this agreement.

I warrant that the compositions sold and assigned to ______________________________, Publisher, hereunder are my original compositions and is not derived from anyone else's work. I further warrant that I have not sold or assigned my interests in such compositions to anyone other than _________________________, Publisher. Interpretation of this agreement shall be according to the laws of the State of ______________. If any portion of this agreement shall be determined invalid or unenforceable, the remainder of the provision shall remain in full force and effect and shall in no way be affected, impaired or invalidated.

A photocopy of this authorization shall have the same force and effect as the original.

Dated: ____________________________

____

Signature

____

Print Name

____

Street

____

City, State, Zip Code

Phone (___)________________________

* Provided by MEDIA BRIDGE TECHNOLOGIES

SAMPLE AUDIOTAPE BUDGET

SAMPLE BUDGET

90 MINUTE SPOKEN-WORD AUDIO CASSETTE -

	Qty	Units	Rate	Total
Studio I				
Record Author	4	HRS	$ 100	$ 400.00
Studio II				
Editing	50	HRS	$ 30	$ 1500.00
Assembly/Mastering	4	HRS	$ 30	$ 120.00
Music Inserts	4	FEE	$ 200	$ 200.00
Producer/Director	1	FEE	$ 500	$ 300.00
Materials	1	FEE	$ 200	$ 200.00
Administrative	1	FEE	$ 50	$ 50.00
Transportation	1	FEE	$ 40	$ 40.00
			TOTAL:	$ 2810.00

* - Studio rates include engineer. 15% Contingency

SAMPLE AUDIOTAPE BUDGET

MEDIA BRIDGE TECHNOLOGIES
Estimated Cost for Story Audio Tapes

20 Minute Audio Cassette (4 Characters) Children

	1 Tape	4 or more (each)
A. Studio Recording		
Record Characters	8	8
Edit Character Voices	16	16
Mix Material and Assemble	10	10
	34	34
Studio Rate $ 35/Hour:	$ 1190.00	$ 1190.00
B. Labor Costs:		
Songwriter*	$ 500.00	$ 500.00
Music Producer/Arranger	$ 800.00	$ 600.00
	$ 1300.00	$ 1100.00
C. Producer Fee	$ 1000.00	$ 600.00
Pre-production		
Production		
Post-production		
D. Material Costs		
1/2" Tape (4)	$ 160.00	$ 160.00
1/4" Tape (2)	$ 50.00	$ 50.00
Audio Cassettes	$ 12.00	$ 12.00
	$ 222.00	$ 222.00
E. Miscellaneous	$ 40.00	$ 40.00
Grand Total:	$ 4197.00	$ 3727.00 (each for 4 or more)

* Composition of four distinct songs or 16 jingles.

15% Contingency.

AUDIOTAPE/CD RESOURCES

HI-SPEED DUPLICATORS
ARP Media
1600 Rich Road
Richmond, IN 47374
(800) 443-2805

Eva-Tone
P.O. Box 7020
Clearwater, FL 33518
(800) 382-8663

Multi-Media
Publishing and Packaging
19360 Rinaldi St. Suite 321
Northridge, CA 91326
(818) 366-7292

Music Annex Duplication
42650 Christy Street
Fremont, CA 94538
(415) 226-0800

Tape Specialties
13411 Saticoy St.
No. Hollywood, CA 91605

REAL TIME DUPLICATION

APEX TAPE Duplication
28 Paul Drive, Suite E
San Rafael, CA 94903
Phone (800) 323-2279

MEDIA PRODUCTION HOUSES
Media Bridge Technologies
138 No. Main Street, Suite 26
Sebastopol, CA 95472
(707) 823-7072
FAX (707) 578-0390

PACKAGERS
Alstrom/Pena Comun.
(Design Studio)
2129 Osuna Road NE
Albuquerque, NM 87113

Apple Printing and Marketing
6033 West Century Blvd., S 110
Los Angeles, CA 90045

Blackburn, Inc.
10150 Crosstown Circle
Eden Praire, MN 55344
(800) 328-6369

Communication Graphics
(Design Studio)
545 Cleveland Ave.
Sebastopol, CA 95472

Cottonwood Printing
2117 Osuna Road
Albuquerque, NM 87113
(800) 777-5341
(505) 345-5341

Davkore (Duplication & Packaging)
1300 Space Park Way
Mountain View, CA 94043
(415) 969-3030

Disc Makers (Duplication & Packaging)
1328 N. 4th St.
Philadelphia, PA 19122
(800) 468-9353

Eva-Tone
P.O. Box 7020
Clearwater, FL 33518
(800) 382-8663

Letter Press Service (Manufacturing)
2704 Girard NE
Alburquerque, NM 87107
(505) 884-0868

Multi Media
19360 Rinaldi St. Suite 321
Northridge, CA 91326
(818) 366-7292

Reliance
225 Belleville Ave.
Bloomfield, NJ 07003
(201) 748-2222

AUDIO EQUIPMENT

Accurate Sound
3515 Edison Way
Menlo Park, CA 94025
(415) 365-2843
(High Speed Duplicators)

AEG
2201-K Fifth Avenue
Ronkonkoma, NY 11779
(516) 467-1200
(High Speed Tape Loader)

APEX Machine
3000 NE 12th Terrace
Fort Lauderdale, FL 33334-4497
(305) 565-APEX
(Audio Cassette Imprinter)

Archon
43 Cherry Street
Danvers, MA 01923
(508) 777-8836
(Audio Cassette Imprinters)

Carver
20121 48th Avenue. West
Lynwood, WA 98046
(206) 775-1202 (Dual Audio Reverse Dubbing Cassette)

Communication Graphics
545 Cleveland Ave.
Sebastopol, CA 95472
(707) 829-3246

Concept Design
1105-A Pomeroy St.
Graham, NC 27253
(919) 229-6500
(Cassette Loaders)

Cottonwood Printing
2117 Osuna Road
Albuquerque, NM 87113

Electro Sound
160 San Gabriel Dr.
Sunnyvale, CA 94086
(408) 245-6600
(Audio cassette Duplicators)

El Mar Plastics
840 E. Walnut St.
Carson, CA 90746
(213) 327-3180
(Cassette Album Boxes)

Full Compass
5618 Odama Road
Madison, WI 53719
(800) 356-5844
(Audio Equipment/Tapes)

Fujii International
18843 Tulsa St.
Northridge, CA 91326
(818) 368-3850

Gauss
9130 Glenoaks Blvd.
Sun Valley, CA 91352
(213) 875-1900
(Audio cassette Duplicators)

KABA
24 Commercial Blvd.
Novato, CA 94949
(800) 231-8273
(415) 883-5222
Duplicators/Imprinters)

OTARI Duplicators
378 Vintage Park Dr.
Foster City, CA 94404
(415) 341-5900
(Audio Cassette Duplicators)

Paktec Automation
28130 Avenue Crocker #308
Valencia, CA 91355
(805) 257-7616
(Audio Cassette Packaging System)

Roldex Industries
9980 Glenoaks Blvd., Suite A
Sun Valley, CA 91352
(818) 504-6294
(Audio Cassette Imprinters)

Sonocraft
360 W. 31 St.,
New York, NY 10001
(212) 760-9300
(Audio Master Cassettes)

Sony Corporation
3 Paragon Dr.
Montvale, NJ 07645-1735
(201) 358-4196
(Audio Cassette Duplicators)

Tascam
7733 Telegraph Rd.
Montebello, CA 90640
(213) 726-0303
(Audio Duplication Equipment)

PRODUCTION SOUND LIBRARIES

Dolby Sound Libraries
100 Potrero Ave.
San Francisco, CA 94103-4813
(415) 558-0200

First Com/Music House
13747 Montfort #220
Dallas, TX 75420
(800) 858-8880

Killer Tracks
6534 Sunset Boulevard
Hollywood, CA 90028
(800) 877-0078
(213) 957-4470 FAX

Production Music Library Assn.
40 East 49th St.
New York, NY 10017

Soper Sound Music Library
P.O. Box 498
Palo Alto, CA 94301
(415) 321-4022

PLASTIC DISPLAY RACKS:

BEEMAX Plastics, Inc.
7424 Santa Monica Boulevard
Los Angeles, CA 90046
(213) 876-1770
(800) 421-4393

Markertek (Video and Audio Supply Catalog)
145 Ulster Ave. Box 397
Saugerties, NY (USA) 12477
Phone (800) 522-2025

AUDIO DISTRIBUTORS/ WHOLESALERS

Audio Publishers Group, Inc.
237 North Michigan St. S-609
South Bend, IN 46601
(800) 234-2726

Baker & Taylor
Audio Catalog
65 E. Main S.,
P.O. Box 6920
Bridgewater, NJ 08807-0920

Bowker Electronic Publishing
On Cassette
245 W. 17th St.
New York, NY 10011

Ingram
347 Reedwood Dr.
Nashville, TN 37217

Inland Book Company
P.O. Box 261
East Haven, CT 06512

Publishers Group West
4065 Hollis St.
Emeryville, CA 94608

Pacific Pipeline
19215 66 Avenue South
Kent, WA 98032-1171

Quality Books
918 Sherwood Dr.
Lake Bluff, IL 60044-2204

BOOK STORE CHAINS

B. Dalton Bookseller
105 Fifth Ave.
New York, NY 10003
(212) 356-7626

Walden Book Company
P.O. Box 10218
Stamford, CT 06904
(203) 356-7626

BOOKS AND DATA

How to Produce a Cassette Program and Never Leave Your Office, William R. Guthy
Cassette Productions Unlimited, Inc.
5796 Martin Rd.
Irwindale, CA 91706
(818) 968-6881

Cassette Talk
Eva-Tone
P.O. Box 7020
Clearwater, FL 33518
(800) EVA-TONE

Media Law For Producers, by Philip Miller
Knowledge Industry Publications, Inc.
701 Westchester Ave.
White Plains, NY 10604

RIGHTS & PERMISSIONS AGENCIES

BZ Rights and Permissions
125 West 72nd St.
New York, NY 10023

Mary Williams Music clearance Corp.
6223 Selma Ave. S-211
Hollywood, CA 90028
(213) 469-3186

The Clearing House
6605 Hollywood Blvd.
Hollywood, CA 90028

Harry Fox Agency
205 East 42nd St.
New York, NY 10017
(212) 370-5330

MARKETING/PUBLIC RELATIONS
(Specialists in Spoken Word Tapes)

Kimberli Munkres
Kimberli & Co.
3550 Round Barn Blvd. S-202
Santa Rosa, CA 95403
(707) 578-0377

Audio Publishers Assoc.
c/o Seth Gerschel
Simon & Schuster Building
1230 Avenue of the Americas
New York, NY 10020
(212) 698-7184

COPYRIGHT REGISTRATION
United States Copyright Office
Register of Copyrights
Library of Congress
Washington, DC 20559
(202) 479-0700 Information
(202) 287-9100 Forms

Form SR (Sound Recording)
Form TX (Books, manuals, computer programs)
Form PA (Multimedia package or music package)

AUDIO REVIEWERS
The Following Is A Sample List Of Audio Reviewers In The Country.

J. C. Martin,
Arizona Daily Star, AZ

Mimi Diamond,
Ashbury Park Press, NJ

Roger Falherty,
Chicago Sun Times

Betty-Lee Fox, editor,
Library Journal

James Cos,
Midwest Book Reviews

Jonathan Gabree,
Newsday

Guy Kelly,
Rocky Mountain News,
Denver, CO

Peter Rowe,
San Diego Union

Voice of America
(Radio Reviews)

Independent Reviewers
(Wire Services)

Bennett Pomerantz
Audioworld
6212 Seabrook Road
Seabrook, MD 20706
(301) 459-1905

Bob Seidenstein
Philadelphia Inquirer
400 N. Broad St.
(215) 854-2432

Late Additions

Packager
Ontae, Inc.
Judy Byers
1660 S. Albion #309
Denver, CO 80222

Audio Equip. & Services & Packager
CAM Audio
2210 Executive Dr.
Garland, TX 75041
(800) 527-3458

INDEX

ORDER FORM

Pathfinder Publishing of California
458 Dorothy Ave.
Ventura, CA 93003
Telephone (805) 642-9278 FAX 805-650-3656

Please send me the following book from Pathfinder Publishing:

_____Copies of ***Let Your Ideas Speak Out* @ $8.95** $_____

Sub-Total $_____
Californians: Please add 6.00% tax. $_____
Shipping* $_____

Grand Total $_____

I understand that I may return the book for a full refund if not satisfied.
Name:____

Address:____

_____________________________ZIP:_______________________

*SHIPPING CHARGES U.S.
Books: Enclose $2.50 for the first book and .50c for each additional book. UPS: Truck; $3.50 for first item, .50c for each additional. UPS Air: $5.00 for first item, $1.50 for each additional item.

If you are interested in receiving Audiotape News and updates of *Let Your Ideas Speak Out* please send us your name and address on the form below..

I am interested in receiving audiotape news and any book updates.

Name:____
Address:____
_____________________________ZIP:_______________________